SpringerBriefs in Statistics

More information about this series at http://www.springer.com/series/8921

Ezra Haber Glenn

Working with the American Community Survey in R

A Guide to Using the acs Package

Springer

Ezra Haber Glenn
Department of Urban Studies and Planning
Massachusetts Institute of Technology
Cambridge, MA, USA

ISSN 2191-544X   ISSN 2191-5458   (electronic)
SpringerBriefs in Statistics
ISBN 978-3-319-45771-0   ISBN 978-3-319-45772-7   (eBook)
DOI 10.1007/978-3-319-45772-7

Library of Congress Control Number: 2016951449

Printed on acid-free paper

This Springer imprint is published by Springer Nature
The registered company is Springer International Publishing AG
The registered company address is: Gewerbestrasse 11, 6330 Cham, Switzerland

Preface

The purpose of this monograph is twofold: first, to familiarize readers with the US Census American Community Survey, including both the potential strengths and the particular challenges of working with this dataset; and second, to introduce them to the acs package in the R statistical language, which provides a range of tools for demographic analysis with special attention to addressing these issues.

In particular, the acs package includes functions to allow users (a) to create custom geographies by combining existing ones provided by the Census, (b) to download and import demographic data from the American Community Survey (ACS) and Decennial Census (SF1/SF3), and (c) to manage, manipulate, analyze, plot, and present this data (including proper statistical techniques for dealing with estimates and standard errors). In addition, the package includes a pair of helpful "lookup" tools, one to help users identify the geographic units they want and the other to identify tables and variables from the ACS for the data they are looking for, and some additional convenience functions for working with Census data.

Acknowledgments

Planners working in the USA all owe a tremendous debt of gratitude to our truly excellent Census Bureau, and this seems as good a place as any to recognize this work. In particular, I have benefited from the excellent guidance the Census has issued on the transition to the ACS: the methodology coded into the acs package draws heavily on these works, especially the *Compass* series cited in the package man pages [7].

I would also like to thank my colleagues in the Department of Urban Studies and Planning at MIT, including Joe Ferreira, Duncan Kincaid, Jinhua Zhao, Mike Foster, and a series of department heads—Larry Vale, Amy Glasmeier, and Eran Ben-Joseph—who have provided consistent and generous support for my work on the acs package and my efforts to introduce programming methods in general—and R in particular—into our Master in City Planning program. Additionally, I am grateful for the graduate students in my "Quantitative Reasoning and Statistical Methods" classes over the years, who have been willing to experiment with R and have provided excellent feedback on the challenges of working with ACS at the local level.

The original coding for the acs package was completed with funding from the Puget Sound Regional Council, working with Public Planning, Research, & Implementation. Portions of this work have been previously presented at the Conference for Computers in Urban Planning and Urban Management (Banff, Alberta, 2011) and the ACS Data Users Conference (Hyattsville, MD, 2015), as well as at workshops and webinars of the Puget Sound Regional Council, the Mel King Institute for Community Building, the Central Massachusetts Regional Planning Agency, and the Orange County R User Group. I am indebted to the organizers and attendees of these sessions for their early input as well as to the excellent R user community and subscribers to the acs users listserv for their ongoing feedback.

Finally, a big thank you to my wife Melissa (for lending me her degree in statistics and public policy) and my children Linus, Tobit, and Mehitabel (for being such strong advocates of open-source software); all four have been patient while I tracked down bugs in the code and helpful as I worked through examples of how we make sense of data.

Cambridge, MA, USA   Ezra Haber Glenn

Contents

Chapter 1
The Dawn of the ACS: The Nature of Estimates

Every 10 years, the U.S. Census Bureau undertakes a complete count of the country's population, or at least attempts to do so; that's what a census is. The information they gather is very limited: this is known as the Census "short form," which consists of only six questions on sex, age, race, and household composition. This paper has nothing to do with that.

Starting in 1940, along with this complete enumeration of the population, the Census Bureau began gathering demographic data on a wide variety of additional topics—everything from income and ethnicity to education and commuting patterns; in 1960 this effort evolved into the "long form" survey, administered to a smaller sample of the population (approximately one in six) and reported in summary files.[1] From that point forward census data was presented in two distinct formats: actual numbers derived from complete counts for some data (the "SF-1" and "SF-2" 100 % counts), and estimates derived from samples for everything else (the "SF-3" tables). For most of this time, however, even the estimates were generally treated as counts by both planners and the general public, and outside of the demographic community not much attention was paid to standard errors and confidence intervals.

Starting as a pilot in 2000, and implemented in earnest by mid-decade, the American Community Survey (ACS) has now replaced the Census long-form survey, and provides almost identical data, but in a very different form. The idea behind the ACS—known as "rolling samples" [1]—is simple: rather than gather a one-in-six sample every 10 years, with no updates in between, why not gather much smaller samples every month on an ongoing basis, and aggregate the results over time to provide samples of similar quality? The benefits include more timely data as well as more care in data collection (and therefore a presumed reduction in non-sampling errors); the downside is that the data no longer represent a single point in time, and the estimates reported are derived from much smaller samples

---

[1]These were originally known as "summary tape files."

© The Author(s) 2016
E.H. Glenn, *Working with the American Community Survey in R*,
SpringerBriefs in Statistics, DOI 10.1007/978-3-319-45772-7_1

(with much larger errors) than the decennial long-form. One commentator describes this situation elegantly as "Warmer (More Current) but Fuzzier (Less Precise)" than the long-form data [6]; another compares the old long-form to a once-in-a-decade "snapshot" and the ACS to a ongoing "video," noting that a video allows the viewer to look at individual "freeze-frames," although they may be lower resolution or too blurry—especially when the subject is moving quickly [2].

To their credit, the Census Bureau has been diligent in calling attention to the changed nature of the numbers they distribute, and now religiously reports margins of error along with all ACS data. Groups such as the National Research Council have also stressed the need to increase attention to the nature of the ACS [5] and in recent years the Census Bureau has increased their training and outreach efforts, including the publication of an excellent series of "Compass" reports to guide data users [7] and additional guidance on their "American FactFinder" website. Unfortunately, the inclusion of all these extra numbers still leaves planners somewhat at a loss as to how to proceed: when the errors were not reported we felt we could ignore them and treat the estimates as counts; now we have all these extra columns in everything we download, without the tools or the perspective to know how to deal with them. To resolve this uncomfortable situation and move to a more productive and honest use of ACS data, we need to take a short detour into the peculiar sort of thing that is an estimate.

1.1 Challenges of Estimates in General

The Peculiar Sort of Thing that is an Estimate Contrary to popular belief, estimates are strange creatures, quite unlike ordinary numbers. As an example, if I count the number of days between now and when a draft of this monograph is due to Springer, I may discover that I have exactly 11 days left to write it: that's an easy number to deal with, whether or not I like the reality it represents. If, on the other hand, I *estimate* that I still have another 6 days of testing to work through before I can write up the last section, then I am dealing with something different: how confident am I that 6 days will be enough? Could the testing take as many as *eight* days? More? Is there any chance it could be done in fewer? (Ha!)

Add to this the complexity of combining *multiple* estimates—for example, if I suspect that "roughly half" of the examples I am developing will need to be checked by a demographer friend, and I also need to complete grading for my class during this same period, which will probably require "around three days of work"—and you begin to appreciate the strange and bizarre ways we need to bend our minds to deal with estimates.

When faced with these issues, people typically do one of two things. The most obvious, of course, is to simply treat estimates like real numbers and ignore the fact that they are really something different. A more epistemologically-honest approach is to think of estimates as "fuzzy numbers," which jibes well with the latest philosophical leanings. Unfortunately, the first of these is simply wrong, and the second is mathematically unproductive. Instead, I prefer to think of estimates as "two-dimensional numbers"—they represent complex little *probability distributions* that spring to life to describe our state of knowledge (or our relative lack thereof). When the estimates are the result of random sampling—as is the case with surveys such as the ACS—these distributions are well understood, and can be easily and efficiently described with just two (or, for samples of small n, three) parameters.

In fact, although the "dimensional" metaphor here may be new, the underlying concept is exactly how statisticians typically treat estimates: we think of *distributions* of maximum likelihood, and describe them in terms of both a center (often confusingly called "the" estimate) and a spread (typically the standard error or margin of error); the former helps us locate the distribution somewhere on the number line and the latter defines the curve around that point. An added advantage of this technique is that it provides a hidden translation (or perhaps a *projection*) from two-dimensions down to a more comfortable one: instead of needing to constantly think about the entire distribution around the point, we are able to use a shorthand, envisioning each estimate as a single point surrounded by the safe embracing brackets of a given confidence interval.

So far so good, until (as noted above), it comes time to *combine* estimates in some way. For this, the underlying mathematics requires that we forego the convenient metaphor of flattened projections and remember that these numbers really do have two-dimensions; to add, subtract, or otherwise manipulate them we must do so up in that 2-D space—quite literally—by squaring the standard errors and working with *variances*. (Of course, once we are done with whatever we wanted to do, we get back down onto the safe flat number line with a dimension-clearing square root.)

Dealing with Estimates in ACS Data All that we have said about estimates in general, of course, applies to the ACS in particular. The ACS provides an unprecedented amount of data of particular value for planners working at the local level, but brings with it certain limitations and added complexities. As a result, when working with these estimates, planners find that a number of otherwise straightforward tasks become quite daunting, especially when one realizes that these problems—and those in the following section on multi-year estimates—can all occur in the same basic operation. (See Sect. 1.4 on page 6.)

In order to *combine* estimates—for example, to aggregate Census tracts into the neighborhoods or to merge sub-categories of variables ("Children under age 5", "Children 5–9 years.", "Children 10–12 years.", etc.) into larger, more meaningful groups—planners must add a series of estimates and *also* calculate the standard error for the sum of these estimates, approximated by the square root of the sum of the squared standard errors for each estimate[2]; the same is true for subtraction, an important fact when calculating t-statistics to compare differences across geography or change over time.[3] A different set of rules applies for multiplying and dividing

---

[2] $SE_{\hat{A}+\hat{B}} \approx \sqrt{SE_{\hat{A}}{}^2 + SE_{\hat{B}}{}^2}$.

[3] $SE_{\hat{A}-\hat{B}} \approx \sqrt{SE_{\hat{A}}{}^2 + SE_{\hat{B}}{}^2}$.

standard errors, with added complications related to how the two estimates are related (one formula for dividing when the numerator is a *subset* of the denominator, as is true for calculating proportions, and a different formula when it is not, for ratios and averages). As a result, even simple arithmetic become complex when dealing with estimates derived from ACS samples.

1.2 Challenges of Multi-Year Estimates in Particular

In addition to these problems involved in using sample estimates and standard errors, the "rolling" nature of the ACS forces local planners to consider a number of additional issues related to the process of deriving estimates from multi-year samples.

Adjusting for Inflation Although it is collected every month on an ongoing basis, ACS data is only reported once a year, in updates to the 1-, 3-, and 5-year products. Internally, figures are adjusted to address seasonal variation before being combined, and then all dollar-value figures are adjusted to represent real dollars *in the latest year of the survey*. Thus, when comparing dollar-value data from the 2006–2008 survey with data from the 2007–2009 survey, users must keep in mind that they are comparing apples to oranges (or at least 2008-priced apples to 2009-priced apples), and the adjustment is not always as intuitive as one might assume: although the big difference between these two surveys would seem to be that one contains data from 2006 and the other contains data from 2009—they both contain the same data from 2007 and 2008—this is not entirely true, since the latter survey has updated *all* the data to be in "2009 dollars." When making comparisons, then, planners must note the end years for both surveys and convert one to the other.

Overlapping Errors Another problem when comparing ACS data across time periods stems from a different aspect of this overlap: looking again at these two three-year surveys (2006–2008 vs. 2007–2009), we may be confronted with a situation in which the data being compared is identical in all ways except for the year (i.e., we are looking at the exact same variables from the exact same geographies). In such a case, the fact that the data from 2007 and 2008 is present in both sets means that we might be *underestimating* the difference between the two if we don't account for this fact: the Census Bureau recommends that the standard error of a difference-of-sample-means be multiplied by $\sqrt{(1 - C)}$, where C represents the percentage of overlapping years in the two samples [7]; in this case, the standard error would thus be corrected by being multiplied by $\sqrt{(1 - \frac{2}{3})} = 0.577$, almost *doubling* the t-statistic of any observed difference.

At the same time, if we are comparing, say, one location or indicator in the first time period with a different location or indicator in the second, this would not be the case, and an adjustment would be inappropriate.

1.3 Additional Issues in Using ACS Data

In addition to those points described about, there are a few other peculiarities in dealing with ACS data, mostly related to "hiccups" with the implementation of the sampling program in the first few years.

Group Quarters Prior to 2006, the ACS did not include group quarters in its sampling procedures.[4] As a result, comparisons between periods that span this time period may under- or over-estimate certain populations. For example, if a particular neighborhood has a large student dormitory, planners may see a large increase in the number of college-age residents—or residents without cars, etc.—when comparing data from 2005 and 2006 (or, say, when comparing data from the 2005–2007 ACS and the 2006–2008 ACS). Unfortunately, there is no simple way to address this problem, other than to be mindful of it.

What Do We Mean by 90 %? Because the ACS reports "90 % margins of error" and not standard errors in raw form, data users must manually convert these figures when they desire confidence intervals of different levels. Luckily, this is not a difficult operation: all it requires is that one divide the given margin of error by the appropriate z-statistic (traditionally 1.645, representing 90 % of the area under a standard normal curve), yielding a standard error, which can be then multiplied by a *different* z-statistic to create a new margin of error.

Unfortunately, in the interest of simplicity, the "90 %" margins of error reported in the early years of the ACS program were actually computed using a z-statistic of 1.65, not 1.645. Although this is not a huge problem, it is recommended that users remember to divide by this different factor when recasting margins of error from 2005 or earlier [7].

The Problem of Medians, Means, Percentages, and Other Non-count Units Another issue that often arises when dealing with ACS data is how to aggregate non-count data, especially when medians, means, or percentages are reported. (Technically speaking, this is a problem related to all summary data, not just ACS estimates, but it springs up in the same place as dealing with standard errors, when planners attempt to combine ACS data from different geographies or to add columns.) The ACS reports both an estimate and a 90 % margin of error for all different types of data, but different types must be dealt with differently. When data is in the form of means, percentages, or proportions—all the results of some prior process of division—the math can become rather tricky, and one really needs to

---

[4]"Group quarters" are defined as "a place where people live or stay, in a group living arrangement, that is owned or managed by an entity or organization providing housing and/or services for the residents.... Group quarters include such places as college residence halls, residential treatment centers, skilled nursing facilities, group homes, military barracks, correctional facilities, and workers' dormitories."

build up the new estimates from the underlying counts; when working with medians, this technically requires second-order statistical estimations of the shapes of the distribution around the estimated medians.

1.4   Putting it All Together: A Brief Example

As a brief example of the complexity involved with these sort of manipulations, consider the following:

> *A planner working in the city of Lawrence, MA, is assembling data on two different neighborhoods, known as the "North Common" district and the "Arlington" district. In order to improve the delivery of translation services for low-income senior citizens in the city, the planner would like to know which of these two neighborhoods has a higher percentage of residents who are age 65 or over and speak English "not well" or "not at all".*

Luckily, the ACS has data on this, available at the census tract level in Table B16004 ("Age By Language Spoken At Home By Ability To Speak English For The Population 5 Years And Over"). For starters, however, the planner will need to combine a few columns—the numerator she wants is the sum of those elderly residents who speak English "not well" and "not at all", and the ACS actually breaks each of these down into four different linguistic sub-categories ("Speak Spanish", "Speak other Indo-European languages", "Speak Asian and Pacific Island languages", and "Speak other languages"). So for each tract she must combine values from $2 \times 4 = 8$ columns—each of which must be treated as a "two-dimensional number" and dealt with accordingly: given the number of tracts, that's $8 \times 3 = 24$ error calculations for each of the two districts.

Once that is done, the next step is to aggregate the data (the combined numerators and also the group totals to be used as denominators) for the three tracts in each district, which again involves working with both estimates and standard errors and the associated rules for combining them: this will require $4 \times 3 = 12$ more error terms. The actual conversion from a numerator (the number of elderly residents in these limited-English categories) and a denominator (the total number of residents in the district) into a proportion involves yet another trick of "two-dimensional" math for each district, yielding—after two more steps—a new estimate with a new standard error.[5] And then finally, the actual test for significance between these two district-level percentages represents one last calculation—a difference of means—to combine these kinds of numbers.

In all, even this simple task required $(24 \times 2) + (12 \times 2) + 2 + 1 = 75$ individual calculations on our estimate-type data, each of which is far more involved than what

---

[5]Note, also, that these steps must be done in the correct order: a novice might first compute the tract-level proportions, and then try to sum or average them, in violation of the points made on page 5 concerning "The Problem of Medians, Means, Percentages, and Other Non-count Units".

would be required to deal with non-estimate numbers. (Note that to compare these numbers with the same data from 2 years earlier to look for significant change would involve the same level of effort all over, with the added complications mentioned on page 4) And while none of this work is particularly difficult—nothing harder than squares and square roots—it can get quite tedious, and the chance of error really increases with the number of steps: in short, this would seem to be an ideal task for a computer rather than a human.

Chapter 2
Getting Started in R

2.1 Introduction

In recent years, the R statistical package has emerged as the leading open-source alternative to applications such as SPSS, Stata, and SAS. In some fields—notably biological modeling and econometrics—R is becoming more widely used than commercial competitors, due in large part to the open source development model which allows researchers to collaboratively design custom-built packages for niche applications. Unfortunately, one area of application that has not been as widely explored—despite the potential for fruitful development—is the use of R for demographic analysis. Prior to the current work, there were a few R packages to bridge the gap between GIS and statistical analysis [4]—and one contribution to help with the downloading and management of spatial data and associated datasets from the 2000 Census [3]—but no R packages existed to manage ACS data or address the types of issues raised above.

Based on a collaborative development model, the acs package is the result of work with local and regional planners, students, and other potential data-users.[1] Through conversations with planning practitioners, observation at conferences and field trainings, and research on both Census resources and local planning efforts that make use of ACS data the author identified a short-list of features for inclusion in the package, including functions to help download, explore, summarize, manipulate, analyze, and present ACS data at the neighborhood scale. The package first launched in beta-stage in 2013, and is currently in version 2.0 (released in March 2016).

In passing, it should be noted that most local planning offices are still a long way from using R for statistical work, whether Census-based or not, and the learning curve is probably too steep to expect much change simply as a result of one new

---

[1]In particular, much of the development of the acs package was undertaken under contract with planners at the Puget Sound Regional Council—see "Acknowledgments" on page v.

© The Author(s) 2016
E.H. Glenn, *Working with the American Community Survey in R*,
SpringerBriefs in Statistics, DOI 10.1007/978-3-319-45772-7_2

package. Nonetheless, one goal in developing acs is that over time, if the R project provides more packages designed for common tasks associated with neighborhood planning, eventually more planners at the margin (or perhaps in larger offices with dedicated data staff) may be willing to make the commitment to learn these tools (and possibly even help develop new ones).

The remainder of this document is devoted to describing how to work with the acs package to download and analyze data from the ACS.

2.2   Getting and Installing R

R is a complete statistical package—actually, a complete programming language with special features for statistical applications—with a syntax and work-flow all its own. Luckily, it is well-documented through a variety of tutorials and manuals, most notably those hosted by the cran project at http://cran.r-project.org/manuals.html. Good starting points include:

- *R Installation and Administration*, to get you started (with chapters for each major operating system); and
- *An Introduction to R*, which provides an introduction to the language and how to use R for doing statistical analysis and graphics.

Beyond these, there are dozens of additional good guides. (For a small sampling, see http://cran.r-project.org/other-docs.html.)

Exact installation instructions vary from one operating system or distribution to the next, but at this point most include an automated installer of one kind or another (a windows .exe installer, a Macintosh .pkg, a Debian apt package, etc.). Once you have the correct version to install, it usually requires little more than double-clicking an installer icon or executing a single command-line function.

Windows users may also want to review the FAQ at http://cran.r-project.org/bin/windows/base/rw-FAQ.html; similarly, Mac users should visit http://cran.r-project.org/bin/macosx/RMacOSX-FAQ.html.

2.3   Getting and Installing the acs Package

2.3.1   Installing from CRAN

The acs package is hosted on the CRAN repository. Once R is installed and started, users may install the package with the install.packages command, which automatically handles dependencies.

```
> # do this once, you never need to do it again
# you may be asked to select a CRAN mirror, and then
# lots of output will scroll past
> install.packages("acs")
--- Please select a CRAN mirror for use in
    this session ---
Loading Tcl/Tk interface ... done
trying URL `http://lib.stat.cmu.edu/R/CRAN/src/contrib/
acs_2.0.tar.gz'Content type `application/x-gzip' length
1437111 bytes (1.4 Mb) opened URL
======================================================
downloaded 1.4 Mb

* installing *source* package `acs' ...
** package `acs' successfully unpacked and MD5 sums
   checked
** R
** data
**  moving datasets to lazyload DB
** inst
** preparing package for lazy loading
Creating a generic function for `summary' from package
   `base' in package `acs'
Creating a new generic function for `apply' in package
   `acs'
Creating a generic function for `plot' from package
   `graphics' in package `acs'
** help
*** installing help indices
** building package indices
** testing if installed package can be loaded

* DONE (acs)

The downloaded source packages are in
          `/tmp/RtmppeCyGj/downloaded_packages'
>
```

After installing, be sure to load the package with library(acs) each time
you start a new session.

```
> # once installed, be sure to load the library:
> library(acs)
```

2.3.2   *Installing from a Zipped Tarball*

If for some reason the latest version of the package in not available through
the CRAN repository (or if, perhaps, you intend to experiment with additional
modifications to the source code), you may obtain the software as a "zipped tarball"
of the complete package. It can be installed just like any other package, although
dependencies must be managed separately. Simply start R and then type:

```
> # do this once, you never need to do it again
> install.packages(pkgs = "acs_2.0.tar.gz",
                    repos = NULL)
* installing *source* package 'acs' ...
** R
** data
**  moving datasets to lazyload DB
** inst
** preparing package for lazy loading
Creating a generic function for 'summary' from package
   'base' in package 'acs'
Creating a new generic function for 'apply' in package
   'acs'
Creating a generic function for 'plot' from package
   'graphics' in package 'acs'
** help
*** installing help indices
** building package indices
** testing if installed package can be loaded

* DONE (acs)

>
```

   (You may need to change the working directory to find the file, or specify a
complete path to the pkgs = argument.) Once installed, don't forget to actually
load the package to make the installed functions available:

```
> # do this every time to start a new session
> library(acs)
Loading required package: stringr
Loading required package: plyr
Loading required package: XML

Attaching package: 'acs'
```

```
The following object(s) are masked from 'package:base':

    apply

>
```

The `acs` package depends on a few other fairly common R packages: `methods`, `stringr`, `plyr`, and `XML`. If these are not already on your system, you may need to install those as well—just use `install.packages("`*package.name*`")`. (Note: when the package is downloaded from the CRAN repository, these dependencies will be managed automatically.)

If installation of the tarball fails, users may need to specify the following additional options (likely for Windows and possibly Mac systems):

```
> install.packages("/path/to/acs_2.0.tar.gz",
                    repos = NULL,
     type = "source")
```

Assuming you were able to do these steps, we're ready to try it out.

2.4 Getting and Installing a Census API Key

To download data via the American Community Survey application program interface (API), users need to request a "key" from the Census. Visit http://api.census.gov/data/key_signup.html and fill out the simple form there, agree to the Terms of Service, and the Census will email you a secret key for only you to use.

When working with the functions described below,[2] this key must be provided as an argument to the function. Rather than expecting you to provide this long key each time, the package includes an `api.key.install()` function, which will take the key and install it on the system as part of the package for all future sessions.

```
> # do this once, you never need to do it again
> api.key.install(key="592bc14cnotarealkey686552b17fda
                  3c89dd389")
>
```

---

[2]Or at least those that require interaction with the API, such as `acs.fetch()`, `acs.lookup()`, and the `check=` option for `geo.make()`.

2.4.1   Using a Blank Key: An Informal Workaround

Currently, the requirement for a key seems to be laxly enforced by the Census API, but is nonetheless coded into the `acs` package. Users without a key may find success by simply installing a blank key (i.e., via `api.key.install(key="")`; similarly, calls to `acs.fetch` and `geo.make(..., check=T)` may succeed with a `key=""` argument. Note that while this may work today, it may fail in the future if the API decides to start enforcing the requirement.

Chapter 3
Working with the New Functions

3.1 Overview

We've tried to make this User Guide as detailed as possible, to help you learn about the many advanced features of the new package. As a result, it may look like there is a lot to learn, but in fact the basics are pretty simple: to get ACS data for your own user-defined geographies, all you need to do is:

1. install and load the package, and (optionally) install an API key (see Sects. 2.3 and 2.4);
2. create a `geo.set` using the `geo.make()` function (see Sect. 3.2);
3. optionally, use the `acs.lookup()` function to explore the variables you may want to download (see Sect. 3.3.3 on page 31);
4. use the `acs.fetch()` function to download data for your new geography (see Sect. 3.3.1 on page 26); and then
5. use the existing functions in the package to work with your data (see worked example in Appendix A and the package documentation).

As a teaser, here you can see one single command that will download ACS data on "Place of Birth for the Foreign-Born Population in the United States" for four Puget Sound counties:

```
> lots.o.data=acs.fetch(geo=geo.make(state="WA",
     county=c(33,35,53,61), tract="*"), endyear=2014,
     table.number="B05006")
```

When I tried this at home, it took about 10 seconds to download—but it's a lot of data to deal with: over 249,000 numbers (estimates and errors for 161 variables for each of a 776 tracts...).

© The Author(s) 2016
E.H. Glenn, *Working with the American Community Survey in R*,
SpringerBriefs in Statistics, DOI 10.1007/978-3-319-45772-7_3

3.2   User-Specific Geographies

3.2.1   Basic Building Blocks: The Single Element geo.set

The geo.make() function is used to create new (user-specified) geographies. At the most basic level, a user specifies some combination of existing census levels (state, county, county subdivision, place, tract, and/or block group), and the function returns a new geo.set object holding this information.[1] If you assign this object to a name, you can keep it for later use. (Remember, by default, functions in R don't save things—they simply evaluate and print the results and move on.)

```
> washington=geo.make(state=53)
> alabama=geo.make(state="Alab")
> yakima=geo.make(state="WA", county="Yakima")
> yakima
An object of class "geo.set"
Slot "geo.list":
[[1]]
"geo" object: [1] "Yakima County, Washington"


Slot "combine":
[1] FALSE

Slot "combine.term":
[1] "aggregate"
```

When specifying the state, county, county subdivision, and/or place, geo.make() will accept either FIPS codes or common names, and will try to match on partial strings; there is also limited support for regular expressions, but by default the searches are case sensitive and matches are expected at the start of names. (For example, geo.make(state="WA", county="Kits") should find Kitsap County, and the more adventurous yakima=geo.make(state="Washi", county=".*kima")   should work to create the same Yakima county geo.set as above.) Important: when creating new geographies, each set of arguments must match with *exactly one* known Census geography: if, for example, the names of two places (or counties, or

---

[1]Note: for reasons that will become clear in a moment, even a single geographic unit—say, one specific tract or county—will be wrapped up as a geo.set. Technically, each individual *element* in the set is known as a geo, but users will rarely (if ever) interact will individual elements such as this; wrapping all groups of geographies—even groups consisting of just one element—in geo.sets like this will help make them easier to deal with as the geographies get more complex. To avoid extra words here, I may occasionally ignore this distinction and refer to user-created geo.sets as "geos."

whatever) would both match, the geo.make() function will return an error.[2] The one exception to this "single match" rule is that for the *smallest* level of geography specified, a user can enter "*" to indicate that *all* geographies at that level should be selected.

tract= and block.group= can only be specified by FIPS code number (or "*" for all); they don't really have names to use. (Tracts should be specified as *six digit* numbers, although initial zeroes may be removed; often *trailing* zeroes are removed in common usage, so a tract referred to as "tract 243" is technically FIPS code 243 00, and "tract 3872.01" becomes 387201.)

When creating new geographies, note, too, that not all combinations are valid[3]; in particular, the package attempts to follow paths through the Census "summary levels" (such as summary level 140: "state-county-tract" or summary level 160: "state-place"). So when specifying, for example, state, county, and place, the county will be ignored.

```
> moxee=geo.make(state="WA", county="Yakima",
  place="Moxee")
Warning message:
In function (state, county, county.subdivision, place,
              tract, block.group)  :
  Using sumlev 160 (state-place)
  Other levels not supported by census api at this time
```

(Despite this warning, the geo.set named moxee was nonetheless created—this is just a warning.)

3.2.2 But Where's the Data...?

Note that these new geo.sets are simply placeholders for geographic entities—they do not actually contain any census data *about* these places. Be patient (or jump ahead to Sect. 3.3 on page 26).

3.2.3 Real geo.sets: Complex Groups and Combinations

OK, so far, so good, but what if we want to create new complex geographies made of more than one known census geography? This is why these things are called

---

[2]This seemed preferable to simply including both matches, since all sorts of place names might match a string, and it is doubtful a user really wants them all.

[3]But don't fret: see Sect. 3.2.7 on page 23.

geo.sets: they are actually *collections* of individual census geographic units, which we will later use to download and manipulate ACS data.

Looking back to when we created the yakima geo.set object (Sect. 3.2.1 on page 16), you can see that the newly created object contained some additional information beyond the name of the place: in particular, all geo.sets include a slot named "combine" (initially set to FALSE) and a slot named "combine.term" (initially set to "aggregate"). When a geo.set consists of just a single geo, these extra slots don't do much, but if a geo.set contains more than one item, these two variables determine whether the geographies are to be treated as a set of individual lines or combined together (and relabeled with the "combine.term").[4] Once we have some more interesting sets, these will come in handy.

To make some more interesting sets, we have a few different options:

Specifying Multiple Geographies through geo.make()   Rather than specifying a single set of FIPS codes or names, a user can pass the geo.make() function *vectors* of any length for state=, county=, and the like. If these vectors are all the same length, they will be combined in sequence; if some are shorter, they will be "recycled" in standard R fashion. (Note that this means if you only specify one item for say, state=, it will be used for all, but if you give two states, they will be alternated in the matching.) For simple combinations, this is probably the easiest way to create sets, but for more complicated things, it can get confusing.

```
> psrc=geo.make(state="WA", county=c(33,35,53,61))
> psrc
An object of class "geo.set"
Slot "geo.list":
[[1]]
"geo" object: [1] "King County, Washington"

[[2]]
"geo" object: [1] "Kitsap County, Washington"

[[3]]
"geo" object: [1] "Pierce County, Washington"

[[4]]
"geo" object: [1] "Snohomish County, Washington"
```

---

[4] All this combining and relabeling takes place when the actual data is downloaded, so up until then you can continue to change and re-change the structure of your geo.sets.

```
Slot "combine":
[1] FALSE

Slot "combine.term":
[1] "aggregate"
```

Adding Existing geo.sets with "+"   If you have already created a few different
geo.sets, you can easily combine them together into a new `geo.set` with the
`"+"` operator. Note that this will create a "flat" `geo.set` (no nesting—see
Sect. 3.2.5 on page 21), regardless of whether the constituent parts are nested
sets.[5]

```
> north.mercer.island=geo.make(state=53, county=33,
     tract=c(24300,24400))
> optional.tract=geo.make(state=53, county=33,
  tract=24500)
> # add in one more tract to create new, larger geo
> north.mercer.island.plus=north.mercer.island +
  optional.tract
> length(north.mercer.island.plus)
[1] 3
> str(north.mercer.island.plus)
Formal class 'geo.set' [package "acs"] with 3 slots
  ..@ geo.list    :List of 3
  .. ..$ :Formal class 'geo' [package "acs"]
         with 3 slots
  .. .. .. ..@ api.for:List of 1
  .. .. .. .. ..$ tract: num 24300
  .. .. .. ..@ api.in :List of 2
  .. .. .. .. ..$ state : num 53
  .. .. .. .. ..$ county: num 33
  .. .. .. ..@ name    : chr "Tract 24300, King County,
                         Washington"
  .. ..$ :Formal class 'geo' [package "acs"] with 3
         slots
  .. .. .. ..@ api.for:List of 1
  .. .. .. .. ..$ tract: num 24400
  .. .. .. ..@ api.in :List of 2
  .. .. .. .. ..$ state : num 53
  .. .. .. .. ..$ county: num 33
```

---

[5]By default, the new set will have combine=FALSE, with one exception: when adding a single-
geography (i.e., length==1) to an existing set with combine= already set to TRUE, the new set
will keep combine=TRUE, essentially "folding in" the new geography.

```
.. .. .. ..@ name    : chr "Tract 24400, King County,
                         Washington"
.. ..$ :Formal class 'geo' [package "acs"]
        with 3 slots
.. .. .. ..@ api.for:List of 1
.. .. .. .. ..$ tract: num 24500
.. .. .. ..@ api.in :List of 2
.. .. .. .. ..$ state : num 53
.. .. .. .. ..$ county: num 33
.. .. .. ..@ name    : chr "Tract 24500, King County,
                         Washington"
..@ combine      : logi FALSE
..@ combine.term: chr "aggregate  +  aggregate"
>
```

Combining geo.sets with "c()" A third way to create new multi-element geo.sets is through the use of R's c() function (short for "combine"). Similar to the way R treats lists with this function, c() will combine geo.sets, but attempt to keep whatever structure they already have in place. The result is often a much more complex kind of nested object. There is real power in this structure, but it can also be a bit tricky; probably best reserved for "power users," but certainly worth playing with. (Hint: try creating different sets and combining them in different ways with c(), and then using length() and str() to examine the results.)

3.2.4 Changing *combine* and *combine.term*

To check the current value of the combine and combine.term slots, you can use the combine() and combine.term() functions; to change these values, simply use combine()= and combine.term=.[6]

```
> combine(north.mercer.island)
[1] FALSE
> combine.term(north.mercer.island)
[1] "aggregate"
> combine(north.mercer.island)=T
> combine.term(north.mercer.island)="North Mercer
 Island"
> north.mercer.island
An object of class "geo.set"
```

---

[6]Or combine() <- and combine.term() <-, for R traditionalists…

```
Slot "geo.list":
[[1]]
"geo" object: [1] "Tract 24300, King County, Washington"

[[2]]
"geo" object: [1] "Tract 24400, King County, Washington"


Slot "combine":
[1] TRUE

Slot "combine.term":
[1] "North Mercer Island"
```

3.2.5 Nested and Flat geo.sets

Remember: by default, the addition operator ("+") will always return "flat" geo.sets, with all the geographies in a single list. The combination operator ("c()"), on the other hand, will generally return nested hierarchies, embedding sets within sets. When working with nested sets like this, the combine flag can be set at each level to aggregate *subsets* within the structure (although be careful—if a higher level of set includes combine=T, you'll never actually see the unaggregated subsets deeper down...).

Using these different techniques, you should be able to create whatever sort of new geographies you want—aggregating some geographies, keeping others distinct (but still bundled as a "set" for convenience), mixing and matching different levels of Census geography, and so on.

Two more helpful shortcuts to keep this all straight:

Setting combine= when creating geo.sets   When creating new user-defined geographies with geo.make(), a user can explicitly set both combine=*new-value* and   combine.term=*new-value* as additional arguments to the function.

flatten.geo.set()   The package also includes a flatten.geo.set() helper function which will iron out even the most complex nested geo.set; it will always return an un-nested geo.set with all the geographies at a single depth, with a length() equal to the number of composite parts.

3.2.6   Subsetting geo.sets

Sometimes, instead of combing geo.sets, users may want to work with just a portion of the an existing set. For this, rather than extending the addition metaphor and developing some sort of "subtraction rule," the package implements methods for R's standard subsetting rules for vectors, using [square brackets].

```
> north.mercer.island[2]
An object of class "geo.set"
Slot "geo.list":
[[1]]
"geo" object: [1] "Tract 24400, King County, Washington"


Slot "combine":
[1] FALSE

Slot "combine.term":
[1] "aggregate (partial)"

> psrc[3:4]
An object of class "geo.set"
Slot "geo.list":
[[1]]
"geo" object: [1] "Pierce County, Washington"

[[2]]
"geo" object: [1] "Snohomish County, Washington"


Slot "combine":
[1] FALSE

Slot "combine.term":
[1] "aggregate (partial)"

>
```

Note that subsetting geo.sets will still always return a complete geo.set, even when selecting only a single geography.

3.2.7   Two Tools to Reduce Frustration in Selecting Geographies

`geo.lookup()`: a Helper to Find What You Need It can often be difficult to find exactly the geography you are looking for, and since (as noted above) `geo.make()` expects single matches to the groups of arguments it is given, this could result in a lot of frustration—especially when trying to find names for places or county subdivisions, which are unfamiliar to many users (and often seem very close or redundant: e.g., knowing whether to look for "Moses Lake city" vs. "Moses Lake CDP"). To help, the package also includes the `geo.lookup()` function, which searches on the same arguments as `geo.make()`, but outputs *all* the matches for your inspection.

Unlike `geo.make()`, `geo.lookup()` looks for matches *anywhere* in the name (except when dealing with state names), and will output a dataframe showing candidates that match some or all of the arguments. (The logic is a little complicated, but basically to be included, a geography *must* match the given state name; when a county and a subdivision are both given, both must match; otherwise, geographies are included that match any—but not necessarily all—of the other arguments.)

```
> geo.lookup(state="WA", county="Ska", county.subdivision="oo")
  state state.name county     county.name county.subdivision
1    53 Washington     NA           <NA>                 NA
2    53 Washington     57   Skagit County                NA
3    53 Washington     59 Skamania County                NA
4    53 Washington     57   Skagit County              92944
5    53 Washington     59 Skamania County              90424
  county.subdivision.name
1                   <NA>
2                   <NA>
3                   <NA>
4       Sedro-Woolley CCD
5     Carson-Underwood CCD
>
> geo.lookup(state="WA", county="Kit", place="Ra")
  state state.name county     county.name place        place.name
1    53 Washington     NA           <NA>    NA              <NA>
2    53 Washington     35   Kitsap County    NA              <NA>
3    53 Washington     37 Kittitas County    NA              <NA>
4    53 Washington     NA   Pierce County 57140   Raft Island CDP
5    53 Washington     NA Thurston County 57220      Rainier city
6    53 Washington     NA     King County 57395   Ravensdale CDP
7    53 Washington     NA  Pacific County 57430      Raymond city
>
```

In the first example, the first row matches just the state (summary level 40); the next two rows show matches at the state and county level (summary level 50); the final two rows show matches that were found looking at state ("WA"), county (containing "Ska"), *and* county subdivision (containing "oo"). In the second example, we see something similar in the first three rows, but after that the rest only match on state-place, ignoring the county selection (like summary level 160), although the county names are included in the output for convenience.

The geo.lookup() function can also accept more than a single string for each argument. In the case of states, the function checks each one independently; in all other cases, matching is done on any and all together (as with a logical "or").[7]

```
> geo.lookup(state=c("WA", "OR"), county=c("M","B"))
     state state.name county        county.name
1       53 Washington     NA               <NA>
2       53 Washington      5      Benton County
3       53 Washington     45       Mason County
4       41     Oregon     NA               <NA>
5       41     Oregon      1       Baker County
6       41     Oregon      3      Benton County
7       41     Oregon     45     Malheur County
8       41     Oregon     47      Marion County
9       41     Oregon     49      Morrow County
10      41     Oregon     51   Multnomah County
>
```

Setting check=T When Using geo.make() Another trick to ensure valid geography matching is to set the check= argument when using geo.make(). When this option is set to TRUE (*not* the default), R will verify each element of the geo.set in turn as it creates it, querying the Census API server. If it encounters an invalid geography, the function will return an error, saving you trouble later; essentially, it helps catch geographies that are technically valid in form but match to no actual census geographies.[8]

```
> no.state=geo.make(state=3) # there is no state with
                             # this FIPS code
An object of class "geo.set"
Slot "geo.list":
[[1]]
"geo" object: character(0)


Slot "combine":
[1] FALSE
```

---

[7]At present, geo.lookup() only accepts and searches on state=, county=, county.subdivision=, and place=; eventually we hope to include lookup support to help find tract and block.group numbers as well.

[8]At present, the function breaks on the first non-match, without a whole lot of help; in the future I'll add in some better error-handling for this.

```
Slot "combine.term":
[1] "aggregate"


> no.state-geo.make(state=3, check=T)
Testing geography item 1:   .... Error in
 file(file, "rt") : cannot
open the connection


> # give it something with a bad county/tract match
> shoreline.nw.border=geo.make(state=53,
    county=c(33, 33, 61, 61, 61),
    tract=c(20100, 20200, 20300, 50600, 50700), check=T,
     combine=T,
    combine.term="Shoreline NW Tracts")
Testing geography item 1: Tract 20100, King County,
 Washington .... OK.
Testing geography item 2: Tract 20200, King County,
 Washington .... OK.
Testing geography item 3: Tract 20300, Snohomish County,
 Washington
 .... Error in file(file, "rt") : cannot open the
 connection
>
> # fix the problem and try again
> shoreline.nw.border=geo.make(state=53,
    county=c(33, 33, 33, 61, 61),
    tract=c(20100, 20200, 20300, 50600, 50700), check=T,
     combine=T,
    combine.term="Shoreline NW Tracts")
Testing geography item 1: Tract 20100, King County,
 Washington .... OK.
Testing geography item 2: Tract 20200, King County,
 Washington .... OK.
Testing geography item 3: Tract 20300, King County,
 Washington .... OK.
Testing geography item 4: Tract 50600, Snohomish County,
 Washington .... OK.
Testing geography item 5: Tract 50700, Snohomish County,
 Washington .... OK.
> shoreline.nw.border
An object of class "geo.set"
Slot "geo.list":
[[1]]
"geo" object: [1] "Tract 20100, King County, Washington"
```

```
[[2]]
"geo" object: [1] "Tract 20200, King County, Washington"

[[3]]
"geo" object: [1] "Tract 20300, King County, Washington"

[[4]]
"geo" object: [1] "Tract 50600, Snohomish County,
 Washington"

[[5]]
"geo" object: [1] "Tract 50700, Snohomish County,
 Washington"


Slot "combine":
[1] TRUE

Slot "combine.term":
[1] "Shoreline NW Tracts"

> # it worked!
> # also, note how we can set combine= and
    combine.term=
> # as arguments to geo.make() -- cool!
```

3.3 Getting Data

Once you've created some geo.sets, you're ready for the fun part: using the package to download data directly from the Census ACS API.[9]

3.3.1 acs.fetch(): The Workhorse Function

Whereas the previous version of the package required users to download data from the Census and then import it into R via the read.acs() function, these steps are

---

[9]Actually, you could download data even without creating a geo.set object first— R's evaluation procedures are perfectly happy letting you use geo.make() "on the fly" and passing the results to the acs.fetch() function: you could enter something like acs.fetch(endyear=2013, geography=geo.make(state="WA", county ="*"), table.number="B01003").

combined in the new `acs.fetch()` function. Assuming you've already installed an API key (see Sect. 2.4 on page 13),[10] the call is quite simple:

```
> # table B01003: "Total Population"
> acs.fetch(geography=psrc, endyear=2011,
           table.number="B01003")
ACS DATA:
 2007--2011 ;
  Estimates w/90% confidence intervals;
  for different intervals, see confint()
                   B01003_001
King County        1908379 +/- 0
Kitsap County      249238 +/- 0
Pierce County      791528 +/- 0
Snohomish County 704536 +/- 0


> # table B05001: "Nativity and Citizenship Status in
 the United States"
> acs.fetch(geography=north.mercer.island.plus,
           endyear=2011,
  table.number="B05001")
ACS DATA:
 2007--2011 ;
  Estimates w/90% confidence intervals;
  for different intervals, see confint()
                   B05001_001    B05001_002    B05001_003
 B05001_004 B05001_005
Census Tract 243 6771 +/- 374 5233 +/- 431 0 +/- 92
 71 +/- 74   896 +/- 225
Census Tract 244 3040 +/- 253 2272 +/- 266 13 +/- 21
 57 +/- 44   311 +/- 91
Census Tract 245 4630 +/- 245 3878 +/- 228 0 +/- 92
 69 +/- 43   483 +/- 137
                   B05001_006
Census Tract 243 571 +/- 177
Census Tract 244 387 +/- 140
Census Tract 245 200 +/- 85
>
```

For each of these geo.sets, `combine=F`, but if we want to get more creative we can try:

```
> combine(north.mercer.island.plus)=T
> combine.term(north.mercer.island.plus)="North Mercer
 Island Tracts"
```

---

[10]And if you haven't, you can simply add a `key=` argument each time.

```
> my.geos=c(psrc, north.mercer.island.plus,
  shoreline.nw.border)
> # table B08013: "Aggregate Travel Time To Work
  (in Minutes) Of Workers By Sex"
> acs.fetch(geo=my.geos, table.number="B08013",
  endyear=2011,
                    col.names=c("Total","Male","Female"))
ACS DATA:
 2007--2011 ;
  Estimates w/90% confidence intervals;
  for different intervals, see confint()
                                Total
King County                     24971250 +/- 189173
Kitsap County              ～    3183505 +/- 83983
Pierce County                   9986285 +/- 116148
Snohomish County                9638070 +/- 109605
North Mercer Island Tracts 118285 +/- 10711.36657948
Shoreline NW Tracts             283540 +/- 19482.3119007986
                                Male
King County                     13972415 +/- 124050
Kitsap County                   1936155 +/- 70636
Pierce County                   5787210 +/- 82246
Snohomish County                5550680 +/- 77512
North Mercer Island Tracts 70055 +/- 8217.77110900517
Shoreline NW Tracts             158090 +/- 15076.8047012621
                                Female
King County                     10998835 +/- 129473
Kitsap County                   1247345 +/- 50974
Pierce County                   4199075 +/- 77010
Snohomish County                4087390 +/- 67221
North Mercer Island Tracts 48235 +/- 6455.83534486436
Shoreline NW Tracts             125450 +/- 11813.2511189765
```

As you can see, when combine=T, acs.fetch will aggregate the data (using the sum method for acs-class objects) when it is downloaded.[11]

Available Data By default, acs.fetch() will download the "Five-Year ACS" (span=5) data from the ACS (dataset="acs"), but these defaults can be

---

[11]Note: At the request of some users, the acs package includes a special one.zero= option for the sum function, which may be desirable when aggregating lots of variables with zero-values for estimates. Since acs.fetch calls sum internally, you can set this option when you call acs.fetch and it will be passed along: for example, one could type   acs.fetch(geo=my.geos, endyear=2011, table.number="B08013", one.zero=T). See help(sum-methods) for more on this.

changed by setting these options to other values.[12] As of version 2.0 of the package, `endyear` is a required option with no default; users must specify the *latest* year for the dataset they are seeking: for example, `endyear=2011` for the 2007–2011 ACS data (or, with `span=3`, for the 2009-2011 data...), or `endyear=2010` for the latest Decennial data (assuming that `dataset="sf1"` or `dataset="sf3"`, of course).

At present, the Census API provides the following, all of which are available using `acs.fetch` with the proper combinations of `endyear`, `span`, and `dataset`:

- American Community Survey (`dataset="acs"`)

  - 5-Year Data: `endyear=` 2009–2014 (i.e., six surveys, 2005–2009 through 2010–2014);
  - 3-Year Data: `endyear=` 2012, 2013;
  - 1-Year Data: `endyear=` 2011, 2012, 2013, 2014.

- Decennial Census Data

  - SF1/Short-Form (`dataset="sf1"`): `endyear=` 1990, 2000, 2010;
  - SF3/Long-Form (`dataset="sf3"`): `endyear=` 1990, 2000.[13]

See http://www.census.gov/data/developers/data-sets.html for more information about available data, including guidance about which geographies are provided for each dataset.

Downloading based on a table number is probably the most fool-proof way to get the data you want, but `acs.fetch()` will also accept a number of other arguments instead of `table.number`. Users can provide strings to search for in table names (e.g., `table.name="Age by Sex"` or `table.name="First Ancestry Reported"`) or keywords to find in the names of variables (e.g., `keyword="Male"` or `keyword="Haiti"`)—but be warned: given how many tables there are, you may get more matches than you expected and suffer from the "download overload" of fast-scrolling screens full of data.[14] On the other hand, if you know you want a specific variable or two (not a whole table, just a few columns of it—such as `variable="B05001_006"` or `variable=c("B16001_058", "B16001_059")`), you can ask for that with `acs.fetch(variable=variable.code, ...)`.

---

[12]Users may set `span=1` or `span=3` for other ACS products, or `span=0` for Decennial data; similarly, use `dataset="sf1"` or `dataset="sf3"` for other census products.

[13]SF3 was discontinued after 2000 and replaced with the ACS.

[14]But don't lose hope: see Sect. 3.3.3 on the `acs.lookup()` tool, which can help with this problem.

3.3.2   More Descriptive Variable Names: `col.names=`

Variable names like B01003_001 and B05001_006 provide a great shorthand, and can be good for experienced users, but most of us prefer something more descriptive. To help, the acs.fetch() function accepts a special argument called col.names, which can take any of the following values:

1. when col.names="auto" (the default), census variable codes are returned;
2. when col.names is given a character vector *the same length as the number of variables in the table*, these names will be used instead as variables for the new acs object; and
3. when col.names="pretty", the function will use descriptive names for the variables (but beware: these can be quite long).

```
> ancestry=acs.fetch(geo=psrc, table.name="People
                      Reporting Ancestry",
     endyear=2011, col.names="pretty")
> ancestry[, 20:30] # just a selection of rows -- it's
 a long table!
ACS DATA:
 2007--2011 ;
   Estimates w/90% confidence intervals;
   for different intervals, see confint()
                     People Reporting Ancestry:   Basque
King County         1125 +/- 267
Kitsap County       41 +/- 34
Pierce County       210 +/- 113
Snohomish County 135 +/- 68
                     People Reporting Ancestry:   Belgian
King County         2928 +/- 466
Kitsap County       428 +/- 196
Pierce County       781 +/- 197
Snohomish County 844 +/- 263
                     People Reporting Ancestry:   Brazilian
King County         1716 +/- 519
Kitsap County       231 +/- 185
Pierce County       124 +/- 91
Snohomish County 221 +/- 97
                     People Reporting Ancestry:   British
King County         17088 +/- 997
Kitsap County       1607 +/- 373
Pierce County       3943 +/- 573
Snohomish County 4735 +/- 599
                     People Reporting Ancestry:   Bulgarian
King County         1659 +/- 409
```

```
Kitsap County      18 +/- 26
Pierce County      213 +/- 123
Snohomish County   444 +/- 248
                   People Reporting Ancestry:  Cajun
King County        234 +/- 141
Kitsap County      49 +/- 41
Pierce County      222 +/- 117
Snohomish County   140 +/- 92
                   People Reporting Ancestry:  Canadian
King County        9996 +/- 984
Kitsap County      1076 +/- 311
Pierce County      3016 +/- 462
Snohomish County   3694 +/- 527
                   People Reporting Ancestry:
 Carpatho Rusyn
King County        49 +/- 38
Kitsap County      3 +/- 4
Pierce County      0 +/- 92
Snohomish County   0 +/- 92
                   People Reporting Ancestry:  Celtic
King County        898 +/- 328
Kitsap County      101 +/- 66
Pierce County      263 +/- 101
Snohomish County   207 +/- 121
                   People Reporting Ancestry:  Croatian
King County        4577 +/- 647
Kitsap County      596 +/- 243
Pierce County      2334 +/- 496
Snohomish County   743 +/- 234
                   People Reporting Ancestry:  Cypriot
King County        0 +/- 92
Kitsap County      0 +/- 92
Pierce County      0 +/- 92
Snohomish County   0 +/- 92
```

3.3.3 The `acs.lookup()` Function: Finding the Variables You Want

Using `acs.fetch()` you can download all the data you need from the Census, provided you either *know* the variable codes or table numbers or are willing to make some educated guesses. This is a fine way to work, and it may be all you need to get started, but for more deliberate users, we've also developed a second lookup tool—known as acs.lookup()—to help identify the tables and variables they are interested in. As with the geo.lookup() tool, the results of acs.lookup() can be named, saved, modified, and eventually passed to acs.fetch() to get data.

Finding the Variables You Want acs.lookup() takes arguments similar to acs.fetch—in particular, table.number, table.name, and keyword, as well as dataset (optional, defaults to dataset="acs"), endyear (required, as with acs.fetch), and span (optional, defaults to span=5)—and searches for matches in the meta-data of the Census tables. When multiple search terms are passed to a given argument (e.g., acs.lookup(endyear=2011, keyword=c("Female", "GED"))), the tool returns matches where *all* of the terms are found; similarly, when more than one lookup argument is used (e.g., acs.lookup(endyear=2011, table.number="B01001", keyword="Female")), the tool searches for matches that include all of these terms (i.e., terms are combined with a logical AND, not a logical OR). Like acs.fetch, string matches with acs.lookup are case sensitive by default, but users may change this by passing case.sensitive=F as an option.

```
> urdu=acs.lookup(keyword="Urdu", endyear=2011)
> urdu
An object of class "acs.lookup"
endyear= 2011   ; span= 5

results:
  variable.code table.number
1    B16001_057       B16001
2    B16001_058       B16001
3    B16001_059       B16001


                                           table.name
1 Language Spoken at Home by Ability to Speak English
  for the Population 5+ Yrs
2 Language Spoken at Home by Ability to Speak English
  for the Population 5+ Yrs
3 Language Spoken at Home by Ability to Speak English
  for the Population 5+ Yrs
                                         variable.name
1                                                 Urdu:
2             Urdu: Speak English 'very well'
3   Urdu: Speak English less than 'very well'

> age.by.sex=acs.lookup(table.name="Age by Sex",
                        endyear=2011)
```

```
> age.by.sex
An object of class "acs.lookup"
endyear= 2011   ; span= 5

results:
  variable.code table.number
 table.name
1     B01002_001       B01002 Median Age by Sex
2     B01002_002       B01002 Median Age by Sex
3     B01002_003       B01002 Median Age by Sex
4     B23013_001       B23013 Median Age by Sex for
 Workers 16 to 64 Years
5     B23013_002       B23013 Median Age by Sex for
 Workers 16 to 64 Years
6     B23013_003       B23013 Median Age by Sex for
 Workers 16 to 64 Years
                variable.name
1  Median age -- Total:
2    Median age -- Male
3  Median age -- Female
4  Median age -- Total:
5      Median age-- Male
6    Median age-- Female
```

Arguments for the acs.lookup are documented in the help files (see ?acs.lookup), but users unfamiliar with ACS variable nomenclature may want to spend a little time testing different search terms, keeping the following in mind:

- The table.number argument is fairly self-explanatory: it usually contains a six-character string, almost always starting with a "B" or "C", followed by a five-digit number (e.g., "B01001" or "C02003"). For tables that include data from Puerto Rico, the table number may include the letters "PR" at the end (e.g., "B05001PR" for "Nativity and Citizenship Status in Puerto Rico"). *Note: For each acs.lookup search, only one table number is allowed.*

- Strings passed to the table.name argument provide search terms to match in the *table names* of the ACS: for example, "Sex" or "Age" or "Age by Sex". Note: these include words that describe *types of categories*, not the categories themselves.

- The keyword argument contains terms to search for in the actual *variable names* of the table. Typically these include descriptive information on the *nominative categories* of the Census on Sex, Age, Race, Language, Ownership, and the like. Examples include "Male", "Female", "Black", "Spanish", "Subsaharan African", "80–84 years", "renter-occupied", and so on. *Note: due to inconsistent capitalization rules, if you don't find the results, you expected, you may want to try again with case.sensitive=F.*

- Don't forget that endyear is a required argument for acs.lookup.
- While dataset and span are optional arguments, variable codes, table numbers, and table names may change from year to year or dataset to dataset, so it's best to specify them as well, just as you would do for acs.fetch.

To help keep it clear, as a rule of thumb: table.name tells you what *sort of categories* the table's variables contain, and keyword tells you what *particular categories* each specific variable includes. So if you want information on *all* races (or age groups or languages, etc.), use table.name="Race" (or "Age" or "Language", etc.); if you only want a specific race (or age group or language, etc.), use keyword="Asian" (and so on).

Manipulating and Using acs.lookup Objects Since acs.lookup objects are valid objects in R, they can be named and saved (for example, urdu and age.by.sex above) and further manipulated by the user. Results contained within acs.objects can be subsetted (with [square brackets]), and even combined (with either c() or +—both function the same way) to create new acs.lookup objects.

```
> workers.age.by.sex=age.by.sex[4:6]
> my.vars=workers.age.by.sex+urdu
> # could also be:
> # my.vars=c(workers.age.by.sex, urdu)
> my.vars
An object of class "acs.lookup"
endyear= 2011   ; span= 5

results:
    variable.code table.number
4       B23013_001      B23013
5       B23013_002      B23013
6       B23013_003      B23013
41      B16001_057      B16001
51      B16001_058      B16001
61      B16001_059      B16001
```

	table.name
4	Median Age by Sex
for Workers 16 to 64 Years	
5	Median Age by Sex
for Workers 16 to 64 Years	
6	Median Age by Sex
for Workers 16 to 64 Years	

```
41 Language Spoken at Home by Ability to Speak English
   for the Population 5+ Yrs
51 Language Spoken at Home by Ability to Speak English
   for the Population 5+ Yrs
61 Language Spoken at Home by Ability to Speak English
   for the Population 5+ Yrs
                                     variable.name
4                           Median age -- Total:
5                            Median age-- Male
6                           Median age-- Female
41                                           Urdu:
51          Urdu: Speak English 'very well'
61  Urdu: Speak English less than 'very well'

> acs.fetch(geography=psrc, endyear=2011,
            variable=my.vars)
ACS DATA:
 2007--2011 ;
  Estimates w/90% confidence intervals;
  for different intervals, see confint()
                           B23013_001    B23013_002
 B23013_003
King County, Washington        39.6 +/- 0.2 39.5 +/- 0.2
 39.6 +/- 0.2
Kitsap County, Washington      41.2 +/- 0.3 40.3 +/- 0.3
 42.3 +/- 0.4
Pierce County, Washington      39.8 +/- 0.2 39.6 +/- 0.2
 40.1 +/- 0.2
Snohomish County, Washington 41.1 +/- 0.2 40.9 +/- 0.2
 41.3 +/- 0.3

                           B16001_057    B16001_058
 B16001_059
King County, Washington        1735 +/- 557 1308 +/- 454
 427 +/- 170
Kitsap County, Washington      75 +/- 99     75 +/- 99
 0 +/- 92
Pierce County, Washington      219 +/- 189  204 +/- 178
 15 +/- 27
Snohomish County, Washington 1179 +/- 520 858 +/- 329
 321 +/- 267
>
```

Note that these "acs.lookup" objects can also be passed as variables to
acs.fetch with different (new) values for endyear and span:

```
> acs.fetch(geography=psrc, endyear=2014,
             variable=my.vars)
ACS DATA:
 2010--2014 ;
  Estimates w/90% confidence intervals;
  for different intervals, see confint()
                                 B23013_001   B23013_002
 B23013_003
King County, Washington          39.7 +/- 0.1 39.8 +/- 0.2
 39.6 +/- 0.2
Kitsap County, Washington        41 +/- 0.3   39.6 +/- 0.4
 42.6 +/- 0.5
Pierce County, Washington        40 +/- 0.2   39.7 +/- 0.2
 40.4 +/- 0.3
Snohomish County, Washington 41.3 +/- 0.2 41.4 +/- 0.2
 41.1 +/- 0.2
                                 B16001_057   B16001_058
 B16001_059
King County, Washington          2343 +/- 527 1880 +/- 425
 463 +/- 168
Kitsap County, Washington        62 +/- 91    62 +/- 91
 0 +/- 28
Pierce County, Washington        108 +/- 161  77 +/- 106
 31 +/- 56
Snohomish County, Washington 1123 +/- 488 860 +/- 318
 263 +/- 211

> acs.fetch(geography=psrc, endyear=2014, span=1,
             variable=my.vars)
ACS DATA:
 2014 ;
  Estimates w/90% confidence intervals;
  for different intervals, see confint()
                                 B23013_001   B23013_002
 B23013_003
King County, Washington          39.4 +/- 0.2 39.6 +/- 0.2
 39.1 +/- 0.4
Kitsap County, Washington        40 +/- 0.7   38.1 +/- 1.4
 41.8 +/- 1.2
Pierce County, Washington        39.3 +/- 0.4 39.2 +/- 0.5
 39.4 +/- 0.7
```

```
Snohomish County, Washington 41 +/- 0.5    41.2 +/- 0.5
 40.7 +/- 0.7
                                B16001_057      B16001_058
 B16001_059
King County, Washington         2844 +/- 1478 2413 +/- 1158
 431 +/- 403
Kitsap County, Washington      NA +/- NA     NA +/- NA
 NA +/- NA
Pierce County, Washington      NA +/- NA     NA +/- NA
 NA +/- NA
Snohomish County, Washington 1856 +/- 1247 1581 +/- 1076
 275 +/- 258
>
```

And, in this way, once the Census has released data for 2015 users may begin to download it *even before the acs package has been updated*:

```
> acs.fetch(geography=psrc, endyear=2015,
 variable=my.vars)
Error in file(file, "rt") : cannot open the connection
> # error now, but when the data is available through
> # the API this will actually work!!
```

Chapter 4
Exporting Data

In the future, versions of the `acs` package will include improved export functions to allow users to save acs data in a variety of formats. For now, however, users wishing to export data for use in spreadsheets or other program can make use of the existing export functions, such as `write.csv`, along with the package's `estimate`, `standard.error`, and `confint` functions. Thus, to save the estimates, standard errors, and a 90 % confidence interval as three different `.csv` spreadsheets:

```
> write.csv(estimate(ancestry),
          file="./ancestry_estimate.csv")
> write.csv(standard.error(ancestry),
          file="./ancestry_error.csv")
> write.csv(confint(ancestry, level=.90),
          file="./ancestry_confint.csv")
```

Depending on the shape you ideally want the data to take, you may want to first create a dataframe from these various elements—a first column of estimate, a second column of 90 % MOEs, for example—and then save that:

```
> urdu.speakers=acs.fetch(geography=c(psrc,
          north.mercer.island.plus),
    variable=urdu[1], endyear=2011,
 col.names="Speak Urdu")
> urdu.speakers
ACS DATA:
 2007--2011 ;
  Estimates w/90% confidence intervals;
  for different intervals, see confint()
                               Speak Urdu
King County, Washington        1735 +/- 557
Kitsap County, Washington      75 +/- 99
```

© The Author(s) 2016

E.H. Glenn, *Working with the American Community Survey in R*,
SpringerBriefs in Statistics, DOI 10.1007/978-3-319-45772-7_4

```
Pierce County, Washington      219 +/- 189
Snohomish County, Washington 1179 +/- 520
North Mercer Island Tracts     0 +/- 159.348674296337

> my.data=data.frame(estimate(urdu.speakers),
      1.645*standard.error(urdu.speakers))
> colnames(my.data)=c("Estimate","90% MOE")
> my.data
                             Estimate  90% MOE
King County, Washington          1735 557.0000
Kitsap County, Washington          75  99.0000
Pierce County, Washington         219 189.0000
Snohomish County, Washington     1179 520.0000
North Mercer Island Tracts          0 159.3487
> write.csv(my.data, file="./urdu.csv")
>
```

Chapter 5
Additional Resources

The acs package is hosted on the CRAN repository, where updates will appear from time to time. For additional guidance and examples, users are advised to review the complete documentation at (http://cran.r-project.org/web/packages/acs/index.html), which can also be accessed in an R session via the help function.

In addition, the "CityState" website http://eglenn.scripts.mit.edu/citystate/ will continue to include updates, patches, worked examples, and more. And finally, users may subscribe to a mailing list at http://mailman.mit.edu/mailman/listinfo/acs-r to keep in touch about the ongoing development of the package, including information on ongoing development; user questions, technical assistance, and new feature requests; and additional updates.

© The Author(s) 2016

E.H. Glenn, *Working with the American Community Survey in R*,

SpringerBriefs in Statistics, DOI 10.1007/978-3-319-45772-7_5

Appendix A
A Worked Example Using Blockgroup-Level Data and Nested Combined geo.sets

To showcase how the package can create new census geographies based on blockgroups—the smallest census geographies provided via the Census API—we can use the following example from Middlesex County in Massachusetts.

A.1 Making the geo.set

To gather data on all the block groups for tract 387201, we create a new geo like this:

```
> my.tract=geo.make(state="MA", county="Middlesex",
   tract=387201, block.group="*", check=T)
Testing geography item 1: Tract 387201, Blockgroup *,
   Middlesex County, Massachusetts .... OK.
>
```

This might be a useful first step, especially if I didn't know how many block groups there were in the tract, or what they were called. Also, note that check=T is not required, but can often help ensure you are dealing with valid geos.

If we then wanted to get very basic info on these block groups—say, table number "B01003" (Total Population), we use:

```
> total.pop=acs.fetch(geo=my.tract, endyear=2011,
                      table.number="B01003")
> total.pop
ACS DATA:
 2007--2011 ;
  Estimates w/90% confidence intervals;
  for different intervals, see confint()
              B01003_001
```

© The Author(s) 2016
E.H. Glenn, *Working with the American Community Survey in R*,
SpringerBriefs in Statistics, DOI 10.1007/978-3-319-45772-7

```
Block Group 1 2681 +/- 319
Block Group 2 952 +/- 213
Block Group 3 1010 +/- 156
Block Group 4 938 +/- 214
>
```

Here we can see that the `block.group="*"` has yielded the actual four block groups for the tract.[1] Now, if instead of wanting all of them, we only wanted the first two, we could just type:

```
> my.bgs=geo.make(state="MA", county="Middlesex",
    tract=387201, block.group=1:2, check=T)
Testing geography item 1: Tract 387201, Blockgroup 1,
  Middlesex County, Massachusetts .... OK.
Testing geography item 2: Tract 387201, Blockgroup 2,
  Middlesex County, Massachusetts .... OK.
>
```

And then:

```
> bg.total.pop=acs.fetch(geo=my.bgs, endyear=2011,
                         table.number="B01003")
> bg.total.pop
ACS DATA:
 2007--2011 ;
  Estimates w/90% confidence intervals;
  for different intervals, see confint()
             B01003_001
Block Group 1 2681 +/- 319
Block Group 2 952 +/- 213
>
```

Now, if we wanted to add in some blockgroups from tract 387100 (a.k.a. "tract 3871"—but remember: we need those trailing zeroes)—say, blockgroups 2 and 3—we could enter:

```
> my.bgs=my.bgs+geo.make(state="MA", county="Middlesex",
    tract=387100, block.group=2:3, check=T)
Testing geography item 1: Tract 387100, Blockgroup 2,
  Middlesex County, Massachusetts .... OK.
Testing geography item 2: Tract 387100, Blockgroup 3,
  Middlesex County, Massachusetts .... OK.
```

---

[1]A similar approach can help find the names of all tracts in a county, for example: `acs.fetch(geography=geo.make(state="MA", county="Middlesex", tract="*"), table.number="B01001")` returns a list of all 300+ tracts in the county, with estimates of total population.

And then:

```
> acs.fetch(geo=my.bgs, endyear=2011,
  table.number="B01003")
ACS DATA:
 2007--2011 ;
  Estimates w/90% confidence intervals;
  for different intervals, see confint()


 B01003_001
Block Group 1, Census Tract 3872.01, Middlesex County,
 ...   2681 +/- 319
Block Group 2, Census Tract 3872.01, Middlesex County,
 ...   952 +/- 213
Block Group 2, Census Tract 3871, Middlesex County,
 ...       827 +/- 171
Block Group 3, Census Tract 3871, Middlesex County,
 ...     1821 +/- 236
>
```

A.2   Using `combine=T` to Make a Neighborhood

Next, to showcase the real power of geo.sets: let's say we don't just want to get data on the four blockgroups, but I want to *combine* them into a single new geographic entity—say, a neighborhood called "Turkey Hill." Before downloading, we could simply say:

```
> combine(my.bgs)=T
> combine.term(my.bgs)="Turkey Hill"'
> acs.fetch(geo=my.bgs, endyear=2011,
  table.number="B01003")
ACS DATA:
 2007--2011 ;
  Estimates w/90% confidence intervals;
  for different intervals, see confint()
            B01003_001
Turkey Hill 6281 +/- 481.733328720362
>
```

And *voila!*, the package sums the estimates and deals with the margins of error, so we don't need to get our hands dirty with square roots and standard errors and all that messy stuff.

A.3   Even More Complex `geo.sets`

We can even create interesting nested geo.sets, where some of the lower levels are combined, and others are kept distinct:

```
> more.bgs=c(my.bgs, geo.make(state="MA",
  county="Middlesex", tract=370300, block.group=1:2, check=T),
  geo.make(state="MA", county="Middlesex", tract=370400,
  block.group=1:3, combine=T, combine.term="Quirky Hill", check=T))
Testing geography item 1: Tract 370300, Blockgroup 1, .... OK.
Testing geography item 2: Tract 370300, Blockgroup 2, .... OK.
Testing geography item 1: Tract 370400, Blockgroup 1, .... OK.
Testing geography item 2: Tract 370400, Blockgroup 2, .... OK.
Testing geography item 3: Tract 370400, Blockgroup 3, .... OK.
> acs.fetch(geo=more.bgs, endyear=2011, table.number="B01003",
  col.names="pretty")
ACS DATA:
 2007--2011 ;
  Estimates w/90% confidence intervals;
  for different intervals, see confint()
                                          Total Population:  Total
Turkey Hill                               6281 +/- 481.733328720362
Block Group 1, Census Tract 3703          315 +/- 132
Block Group 2, Census Tract 3703          1460 +/- 358
Quirky Hill                               2594 +/- 487.719181496894
>
```

We can even create a `geo.set` that bundles different levels of census geography—for example, our two neighborhoods ("Turkey Hill" and "Quirky Hill"), plus some data for comparison on the entire county and state level.

```
> neighborhood.geos=c(more.bgs[c(1,3)],
  geo.make(state="MA", county="Middlesex"),
  geo.make(state="MA"))
> acs.fetch(geography=neighborhood.geos, endyear=2011,
  table.number="B01003", col.names="pretty")
ACS DATA:
 2007--2011 ;
  Estimates w/90% confidence intervals;
  for different intervals, see confint()
                                          Total Population: Total
Turkey Hill                               6281 +/- 481.733328720362
Quirky Hill                               2594 +/- 487.719181496894
Middlesex County, Massachusetts 1491762 +/- 0
Massachusetts                             6512227 +/- 0
>
```

Note that this geo.set can now be used again and again to download and analyze many different variables for these same geographies.

A.4   Gathering Neighborhood Data on Transit Mode-Share

As a final example, let's look for some data on commuting choices for these two neighborhoods, compared to the county and state. If we don't know what census variables we wants, we can use the acs.lookup function to search for likely candidates. Let's see which variables use the word "Bicycle":

```
> acs.lookup(keyword="Bicycle", endyear=2011)
An object of class "acs.lookup"
endyear= 2011   ; span= 5

results:
   variable.code table.number
1      B08006_014       B08006
2      B08006_031       B08006
3      B08006_048       B08006
4      B08301_018       B08301
5      B08406_014       B08406
6      B08406_031       B08406
7      B08406_048       B08406
8      B08601_018       B08601


                                              table.name
1                             Sex of Workers by Means of
   Transportation to Work
2                             Sex of Workers by Means of
   Transportation to Work
3                             Sex of Workers by Means of
   Transportation to Work
4                                               Means of
   Transportation to Work
5 Sex of Workers by Means of Transportation to Work
   for Workplace Geography
6 Sex of Workers by Means of Transportation to Work
   for Workplace Geography
7 Sex of Workers by Means of Transportation to Work
   for Workplace Geography
8                   Means of Transportation to Work
   for Workplace Geography
         variable.name
1             Bicycle
2      Male: Bicycle
3   Female: Bicycle
4             Bicycle
5             Bicycle
```

```
6     Male: Bicycle
7   Female: Bicycle
8           Bicycle

>
```

We've quickly narrowed a few thousand variables down to just 8. As is common with the ACS, there are a number of tables that relate to the topic we are interested in (means of transportation), often cross-tabulated with other topics. The simplest one seems to be the fourth in the list, "Means of Transportation to Work," from table number B08301. Let's look at all the variables there, just to be sure:

```
> acs.lookup(table.number="B08301", endyear=2011)
An object of class "acs.lookup"
endyear= 2011  ; span= 5

results:
    variable.code table.number                        table.name
1     B08301_001      B08301 Means of Transportation to Work
2     B08301_002      B08301 Means of Transportation to Work
3     B08301_003      B08301 Means of Transportation to Work
4     B08301_004      B08301 Means of Transportation to Work
5     B08301_005      B08301 Means of Transportation to Work
... [abbreviated for space]
                                               variable.name
1                                                     Total:
2                                        Car, truck, or van:
3                            Car, truck, or van: Drove alone
4                              Car, truck, or van: Carpooled:
5   Car, truck, or van: Carpooled: In 2-person carpool
... [abbreviated for space]

>
```

This seems to be what we want, including data on people who drove to work alone, biked, took public transit, and so on for 20 different modes (as well as the all important "Total" on the first line, which we will need for percentages). For our purposes, let's look at just a few of these variables: drove alone, public transportation, biking, and the total population from the table.[2] We can subset these and save them as a new acs.lookup object, and pass them right on to fetch some data.

```
> transit.vars=acs.lookup(table.number="B08301")[c(1,3,10,18),
    endyear=2011]
> transit.vars
```

---

[2]Note the importance of the last of these variables: when computing percentages for ACS data, always use the totals from the particular table, not from some other "Total population" table.

```
An object of class "acs.lookup"
endyear= 2011   ; span= 5

results:
   variable.code table.number                              table.name
1      B08301_001      B08301 Means of Transportation to Work
3      B08301_003      B08301 Means of Transportation to Work
10     B08301_010      B08301 Means of Transportation to Work
18     B08301_018      B08301 Means of Transportation to Work
                                          variable.name
1                                             Total:
3                        Car, truck, or van: Drove alone
10   Public transportation (excluding taxicab):
18                                           Bicycle
```

```
> transit.data=acs.fetch(geography=neighborhood.geos,
  variable=transit.vars, endyear=2011,
  col.names=c("Total","Drove Alone","Public Transit","Biked"))
> transit.data
ACS DATA:
 2007--2011 ;
  Estimates w/90% confidence intervals;
  for different intervals, see confint()
                                      Total
Turkey Hill                           3159 +/- 405.076535978079
Quirky Hill                           1891 +/- 380.596899619532
Middlesex County, Massachusetts 773894 +/- 3339
Massachusetts                         3202521 +/- 8062
                                      Drove Alone
Turkey Hill                           2687 +/- 352.326553072572
Quirky Hill                           1068 +/- 301.584150777192
Middlesex County, Massachusetts 539042 +/- 3602
Massachusetts                         2316985 +/- 8271
                                      Public Transit
Turkey Hill                           110 +/- 133.285408053545
Quirky Hill                           333 +/- 133.007518584477
Middlesex County, Massachusetts 82883 +/- 1931
Massachusetts                         291160 +/- 3799
                                      Biked
Turkey Hill                           0 +/- 190
Quirky Hill                           40 +/- 103.009708280336
Middlesex County, Massachusetts 9661 +/- 725
Massachusetts                         21938 +/- 1195
>
```

Since these are raw counts, and we might be more interested in percentages, we can use the special math functions of the acs package to divide the last three columns by the first. (The division function will automatically deal with both estimates and standard errors.) In some cases, division on acs objects is quite simple: something like `transit.data[,2]/transit.data[,1]` would convert the second column from counts to percentages. We can try that here, as follows:

```
> transit.data[,2]/transit.data[,1]
ACS DATA:
 2007--2011 ;
  Estimates w/90% confidence intervals;
  for different intervals, see confint()
                               ( Drove Alone :
 Total )
Turkey Hill                    0.850585628363406 +/-
 0.155998230670757
Quirky Hill                    0.564780539397144 +/-
 0.195848029196112
Middlesex County, Massachusetts 0.696532083205193 +/-
 0.00554027354078655
Massachusetts                  0.723487839736258 +/-
 0.00316025915803343
Warning message:
In transit.data[, 2]/transit.data[, 1] :
  ** using the more conservative formula for
  ratio-type dividing, which does not assume that
  numerator is a subset of denominator; for more
  precise results when seeking a proportion and not a
  ratio, use divide.acs(..., method="proportion") **
>
```

In this case, however, as the warning notes, this is actually slightly wrong: since this should in fact be a "proportion-type" division (and not a "ratio-type" division—see `?divide.acs`), we don't want standard division with `"/"`, but instead must use the package's special `acs.divide` function. This can be called on each column of our data with R's standard `apply` function, which has been adapted to work on acs objects.

```
> apply(transit.data[,2:4], MARGIN=1, FUN=divide.acs,
  denominator=transit.data[,1], method="proportion",
  verbose=F)
ACS DATA:
 2007--2011 ;
  Estimates w/90% confidence intervals;
  for different intervals, see confint()
                               ( Drove Alone /
                                 Total )
Turkey Hill                    0.850585628363406 +/-
 0.0233001603324679
```

```
Quirky Hill                       0.564780539397144 +/-
   0.111865144361133
Middlesex County, Massachusetts 0.696532083205193 +/-
   0.00355414596064069
Massachusetts                     0.723487839736258 +/-
   0.00183110720072149
                                  ( Public Transit /
                                  Total )
Turkey Hill                       0.034821145932257 +/-
   0.0419553490713477
Quirky Hill                       0.176097303014278 +/-
   0.0607546663302448
Middlesex County, Massachusetts 0.107098646584674 +/-
   0.00245201395451528
Massachusetts                     0.0909158753369611
   +/- 0.00116396486009957
                                  ( Biked / Total )
Turkey Hill                       0 +/-
   0.0601456157011713
Quirky Hill                       0.0211528291909043
   +/- 0.0546397830403873
Middlesex County, Massachusetts 0.0124836217879968
   +/- 0.000938367861201701
Massachusetts                     0.00685022830451385
   +/- 0.000373541799644013
>
```

Note in passing that the resulting *estimates* are the same as in the previous division, but that there errors are slightly different as a result of the proportion-type operation.[3]

Now we can see something interesting in our data: not only do far more people in Turkey Hill drive alone (and far fewer take public transit) than in Quirky Hill (or even in the county or state), the differences seem far beyond the report margin of errors.

---

[3]If you don't set verbose=F, the function also returns some warnings—the first two just to let you know that proportion-division is not the same as ratio-division; the third lets you know that in one case, the function defaulted to ratio-style division as per Census guidance.

References

1. Alexander, C.H.: Still rolling: Leslie Kish's "Rolling Samples" and the American Community Survey. In: Proceedings of Statistics Canada Symposium (2001)
2. Alexander, C.H.: A discussion of the quality of estimates from the American Community Survey for small population groups. Technical Report. U.S. Census Bureau, Washington, DC (2002)
3. Almquist, Z.W.: US Census spatial and demographic data in R: the UScensus2000 suite of packages. J. Stat. Softw. **37**(6), 1–31 (2010)
4. Bivand, R.S., Pebesma, E.J., Gómez-Rubio, E.J.: Applied Spatial Data Analysis with R. Springer, New York (2008)
5. Citro, C.F., Kalton, G. (eds.): Using the American Community Survey: Benefits and Challenges. National Research Council, Washington, DC (2007)
6. MacDonald, H.: The American Community Survey: warmer (more current), but fuzzier (less precise) than the decennial census. J. Am. Plan. Assoc. **72**(4), 491–504 (2006)
7. U.S. Census Bureau: A compass for understanding and using American Community Survey data: What state and local governments need to know. Technical Report, U.S. Census Bureau, Washington, DC (2009)

© The Author(s) 2016
E.H. Glenn, *Working with the American Community Survey in R*,
SpringerBriefs in Statistics, DOI 10.1007/978-3-319-45772-7

Printed in the United States
By Bookmasters